Let's Learn About Space

THE SUN

Rebecca Kraft Rector

Enslow Publishing
101 W. 23rd Street
Suite 240
New York, NY 10011
USA
enslow.com

Words to Know

atmosphere Air that surrounds a planet.

core The center of an object.

eclipse When an object is hidden by another.

gravity The force that attracts two bodies to each other.

nuclear fusion A process where a lot of energy is given off when atoms are joined.

orbit The path of an object as it circles another object in space.

rotate To turn in a circle.

solar system The sun and the bodies that move around it.

CONTENTS

The sun is a giant ball of gases.

What Is the Sun?

The sun is a star. It is made of hot gas. The gas is mostly hydrogen. There is some helium, too. Gravity pulls the gas into a ball.

Fast Fact

The sun is a yellow dwarf star.

Planets, asteroids, comets, and more all orbit around the sun.

The Sun's Place in Space

The sun is in the center of our solar system. It is the biggest and brightest object. The sun's gravity pulls objects into orbits around it.

FAST FACT

Eight planets orbit the sun.

A solar eclipse happens several times a year.

Solar Eclipse

Sometimes the moon moves in front of the sun. Sunlight is blocked. This is called a solar eclipse. The sky gets dark for a few minutes.

Different parts of the sun rotate at different speeds.

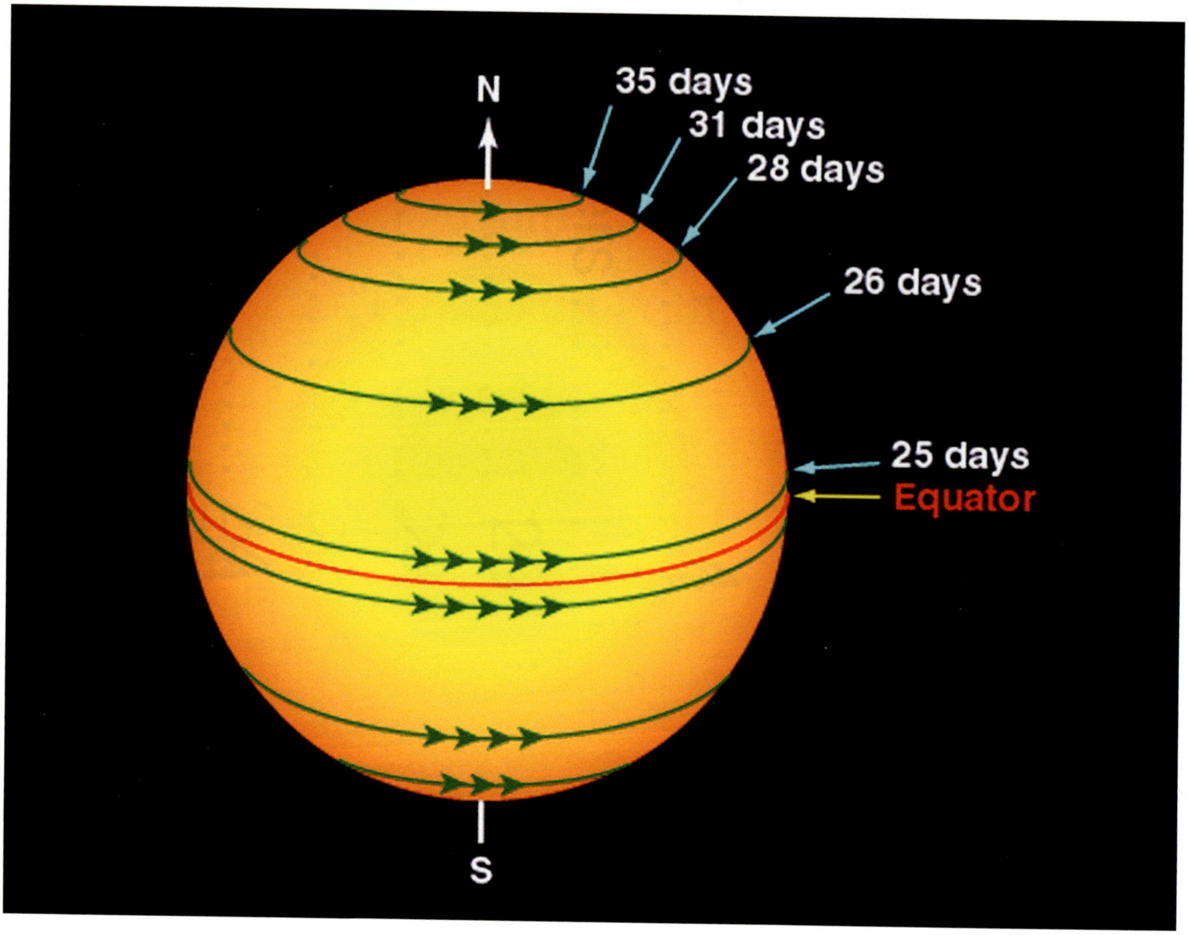

Spinning Gases

The sun **rotates**. The gases spin at different speeds. The middle takes 25 days to make one turn. The top and bottom take 36 days.

Fast Fact

The sun is about 865,000 miles (1,392,000 kilometers) across.

People, animals, and plants all need the sun in order to live.

Energy for Life

The sun gives off energy. Some of it is light and heat. This energy comes to Earth. This is important. All life needs the sun's energy.

FAST FACT

The sun is more than 4.6 billion years old.

The center of the sun is called the core. It is the hottest part of the sun.

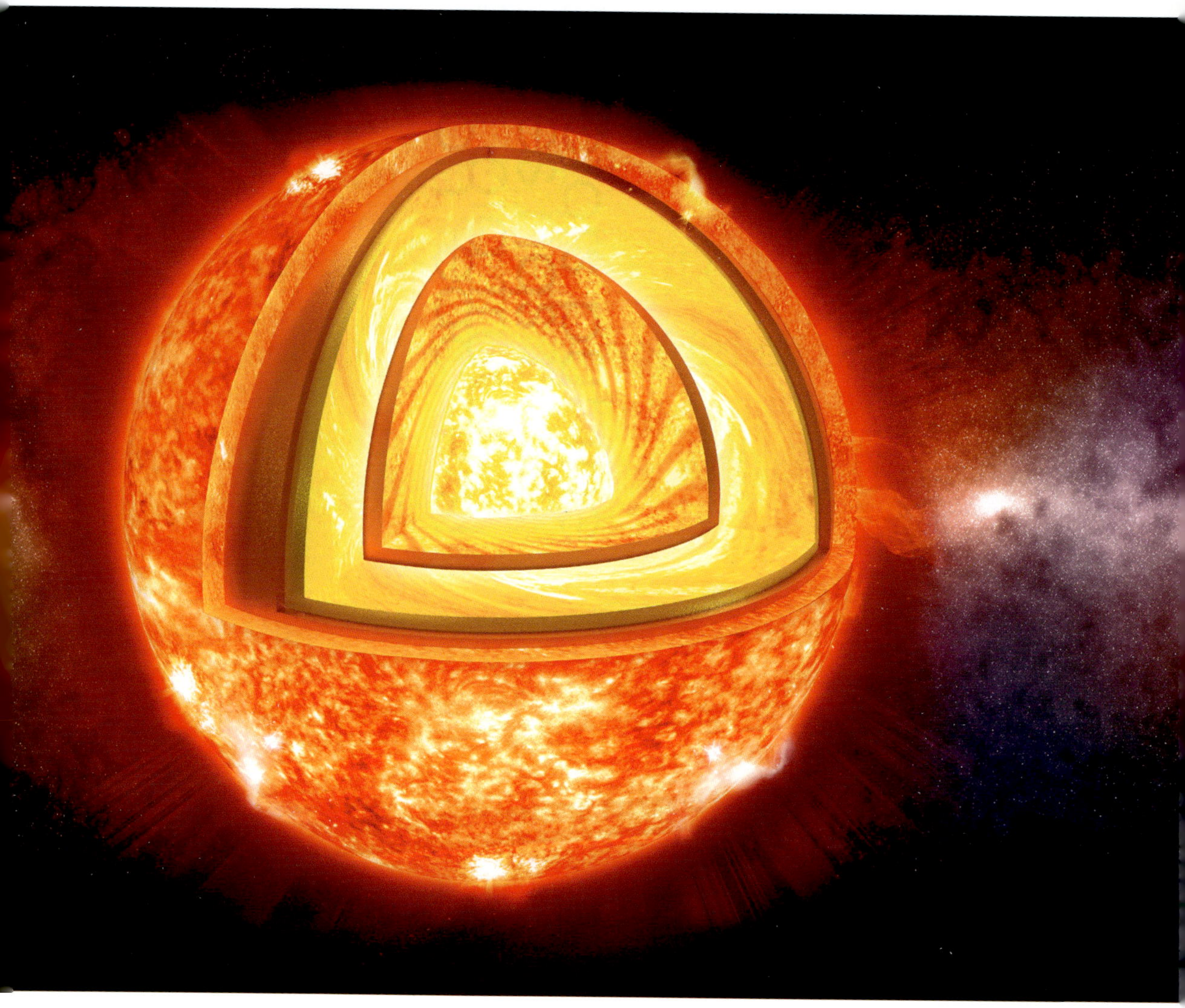

The Core

The sun's core is very hot. It is more than 28 million degrees Fahrenheit (15.5 million degrees Celsius). It is where nuclear fusion changes hydrogen into helium. The change makes energy. The energy spreads into space.

The darker areas you see on the sun are called sunspots.

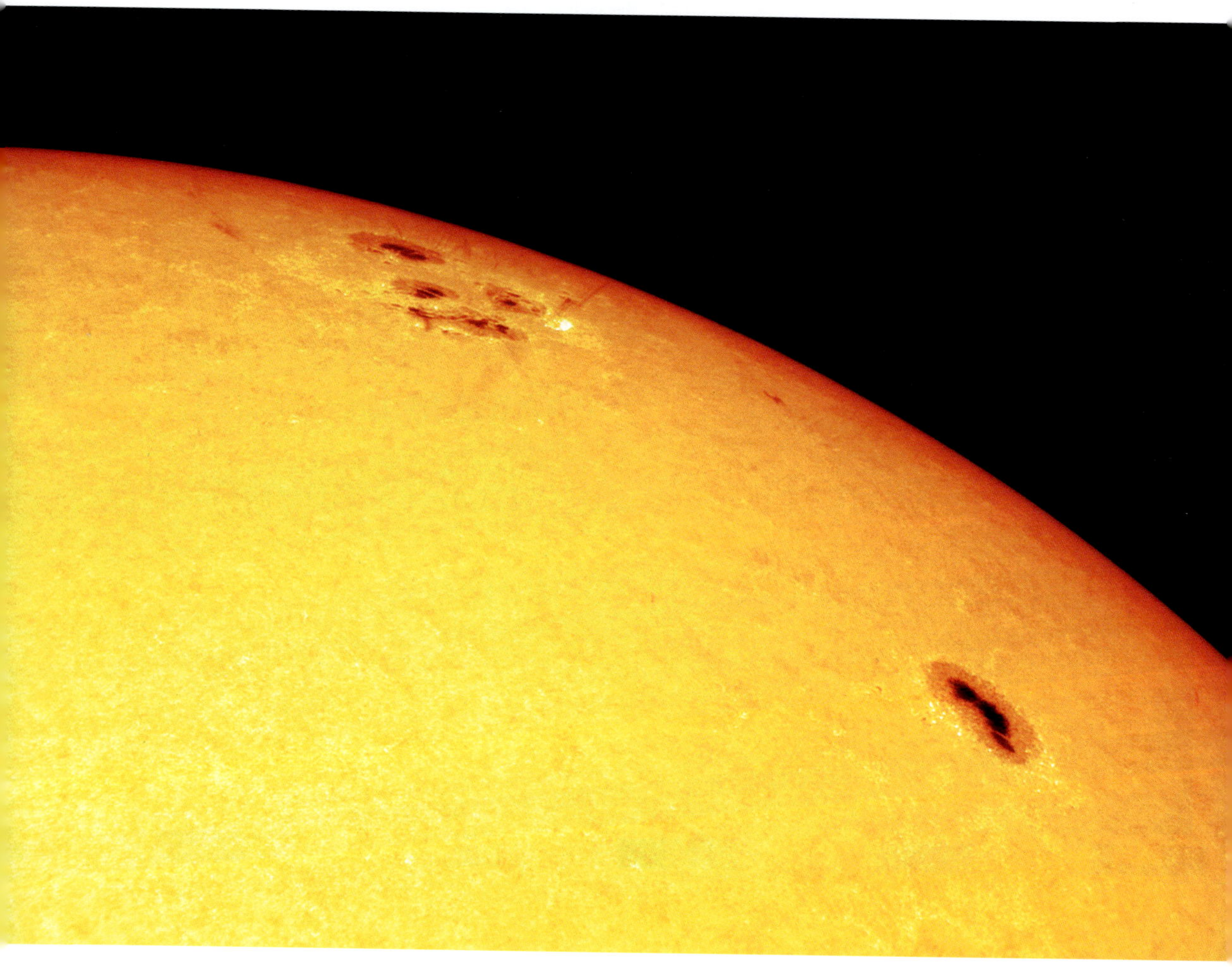

The Surface

The surface of the sun is gas. Sometimes dark spots show up. They are called sunspots. They are storms. They are cooler than the rest of the surface.

Fast Fact

The sun's surface is about 10,000°F (5,500°C).

The corona is the streaks that you see around the sun. It reaches millions of miles out into space.

The Corona

The corona is a layer of gas. It is above the sun's surface. It makes a halo around the sun. We can see it during a total solar eclipse.

FAST FACT

The corona is hundreds of times hotter than the surface of the sun.

Auroras are also known as the northern lights. The colorful glow can usually be seen close to Earth's poles.

Solar Winds

Solar winds are streams of gas. They come from the corona. They can hit Earth's atmosphere. This makes a glow. The glow is called an aurora.

Activity

Me and My Shadow

Materials

A partner
Pavement
Sidewalk chalk
A sunny day

Earth is rotating. This makes the sun seem to move across the sky. Does this change your shadow?

Let's get started!

Procedure:

Step 1: In the morning, draw an X on the pavement.

Step 2: One person stands on the X.

Step 3: The partner traces the other person's shadow with chalk.

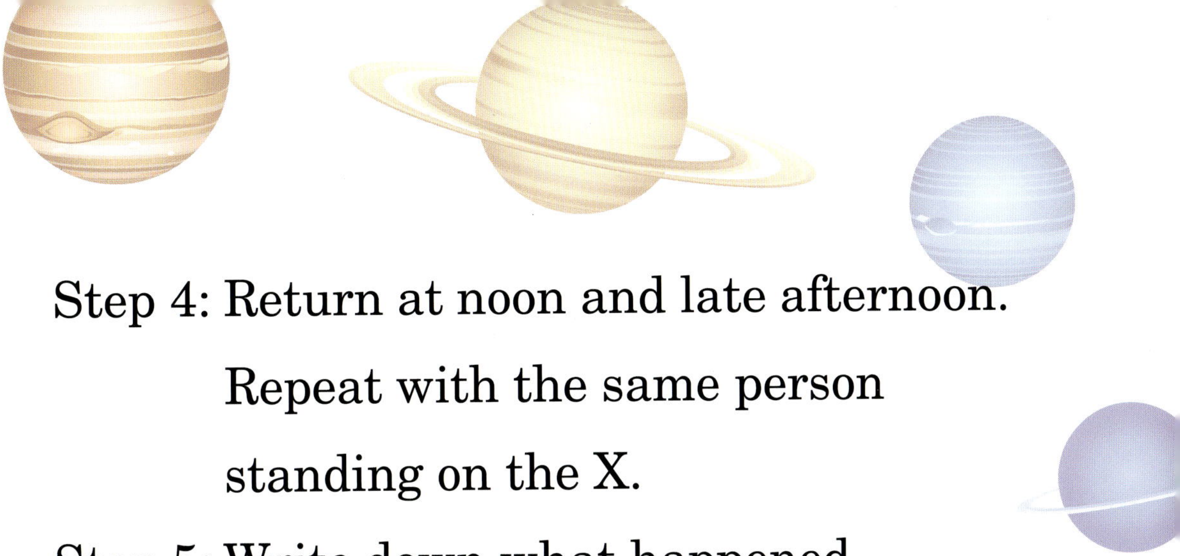

Step 4: Return at noon and late afternoon. Repeat with the same person standing on the X.

Step 5: Write down what happened.

Did your shadow change? Why did that happen?

Use your shadow to trace the position of the sun in the sky.

Learn More

Books

Crane, Cody. *The Sun*. New York, NY: Children's Press, 2018.

McAnulty, Stacy. *Sun! One in a Billion*. New York, NY: Henry Holt, 2018.

Seluk, Nick. *The Sun Is Kind of a Big Deal*. New York, NY: Orchard Books, 2018.

Websites

Kids Astronomy
kidsastronomy.com/solar-system/fun-facts-about-our-sun
Enjoy fun facts about the sun.

NASA Space Place
spaceplace.nasa.gov/sunburn/en
Discover why the sun burns us.

NASA—The Sun
starchild.gsfc.nasa.gov/docs/StarChild/solar_system_level1/sun.html
Check out facts about the sun.

Index

Published in 2020 by Enslow Publishing, LLC.
101 W. 23rd Street, Suite 240, New York, NY 10011

Library of Congress Cataloging-in-Publication Data

Names: Rector, Rebecca Kraft, author.
Title: The sun / Rebecca Kraft Rector.
Description: New York : Enslow Publishing, 2020. | Series: Let's learn about space | Audience: Grade K to 4. | Includes bibliographical references and index.
Identifiers: LCCN 2018042016| ISBN 9781978507333 (library bound) | ISBN 9781978509405 (pbk.) | ISBN 9781978509429 (6 pack) Subjects: LCSH: Sun—Juvenile literature. Classification: LCC QB521.5 .R43 2019 | DDC 523.7—dc23
LC record available at https://lccn.loc.gov/2018042016

Printed in the United States of America

To Our Readers: We have done our best to make sure all website addresses in this book were active and appropriate when we went to press. However, the author and the publisher have no control over and assume no liability for the material available on those websites or on any websites they may link to. Any comments or suggestions can be sent by e-mail to customerservice@enslow.com.

Photos Credits: Cover, p. 1 Igor Kovalchuk/Shutterstock.com; p. 4 Aphelleon/Shutterstock.com; p. 6 Microstocker.Pro/Shutterstock.com; p. 8 NASA/Aubrey Gemignani; p. 10 NASA; p. 12 UvGroup/Shutterstock.com; p. 14 © iStockphoto.com/adventtr; p. 16 Roman Vanur/Shutterstock.com; p. 18 Mark Garlick/Science Photo Library/Getty Images; p. 20 Simon'sPassion4Travel/Shutterstock.com; p. 23 Lorenzo Dominguez/Moment/Getty Images; interior design elements (planets) Vectomart/Shutterstock.com, (sun) Kirill Kirsanov/Shutterstock.com.